TECHNIQUES TO LEARN & REMEMBER MULTIPLICATION TABLES 11 to 20

[An interesting and active alternative method to rote memorization]

By

JOHN LOUIS

Grand Master of Memory

Indian National Memory Champion

International Memory Coach

Human Calendar & Mental Calculator

Teacher

A great exercise for mental muscles to enhance working memory capacity

CONTENT

Author's Note

Dear Teachers and Parents,

Once the children master all the single digit multiplication tables, it becomes so easy to learn the multiplication tables 11 to 20. It involves just 2 simple steps for all the tables. These methods create a pathway to think and derive the answer for every facts of multiplication. Although it might take time to understand and get used to it initially, it becomes very easy to derive the answer from nowhere as quickly as rote memorisation. There is no way for anyone to forget the multiplication facts.

These methods help the students to improve their working memory that is very important for any learning. Moreover, the process of deriving the answers mentally is enjoyable and fun.

I was a chemistry teacher in a school and noticed that the students took more time to solve very simple chemistry problems that just had the arithmetic operation of multiplication and division.

After watching them closely for few weeks, I was shocked to realize the issue went well beyond the current mathematical problem. They did not know how to multiply without counting it out. Somehow the multiplication tables they should have learned in their primary school previously had evaded them, and they were essentially struggling in math.

However, I had a wonderful maths teacher in my grade 6 to 8 in the early 1970's. He showed us how multiplication tables can be mastered in a non-mathematical way. Although, I had learned all the multiplication tables very thoroughly in my primary school itself by constant repetition and drilling, the tricks he showed us for the struggling students was very interesting to me. It made me to think and derive the answer for every simple multiplication. In few weeks of time, even the already struggling students were able to catch up with me and started doing equally well and sometime I had to lose to them! Hence, I like to share those tricks with you here.

Teach one table at a time in the order they are given in this book. Practice each table on its own until it is mastered.

John Louis

Grand Master of Memory
Indian National Memory Champion [2 times]
International Memory Coach
Human Calendar & Mental Calculator
Teacher

CHAPTER-1 : Prior Knowledge

1. Multiplication Tables 1 to 10
2. Counting of numbers from 1 to 200
3. Mental addition of two single digit numbers

Multiplication Tables 1 to 10

Please refer to volume-I of this series of books – Techniques to learn and remember multiplication tables 1 to 10

Counting of numbers from 1 to 100

1	11	21	31	41	51	61	71	81	91
2	12	22	32	42	52	62	72	82	92
3	13	23	33	43	53	63	73	83	93
4	14	24	34	44	54	64	74	84	94
5	15	25	35	45	55	65	75	85	95
6	16	26	36	46	56	66	76	86	96
7	17	27	37	47	57	67	77	87	97
8	18	28	38	48	58	68	78	88	98
9	19	29	39	49	59	69	79	89	99
10	20	30	40	50	60	70	80	90	100

Counting of numbers from 101 to 200

101	111	121	131	141	151	161	171	181	191
102	112	122	132	142	152	162	172	182	192
103	113	123	133	143	153	163	173	183	193
104	114	124	134	144	154	164	174	184	194
105	115	125	135	145	155	165	175	185	195
106	116	126	136	146	156	166	176	186	196
107	117	127	137	147	157	167	177	187	197
108	118	128	138	148	158	168	178	188	198
109	119	129	139	149	159	169	179	189	199
110	120	130	140	150	160	170	180	190	200

Mental addition of two single digit numbers [1 to 9]

1+ 1	1+ 2	1+ 3	1+ 4	1+ 5
1	3	4	5	6
1+ 6	1+ 7	1+ 8	1+ 9	
7	8	9	10	

2+ 1	2+ 2	2+ 3	2+ 4	2+ 5
3	4	5	6	7
2+ 6	2+ 7	2+ 8	2+ 9	
8	9	10	11	

3+ 1	3+ 2	3+ 3	3+ 4	3+ 5
4	5	6	7	8
3+ 6	3+ 7	3+ 8	3+ 9	
9	10	11	12	

4+ 1	4+ 2	4+ 3	4+ 4	4+ 5
5	6	7	8	9
4+ 6	4+ 7	4+ 8	4+ 9	
10	11	12	13	

5+ 1	5+ 2	5+ 3	5+ 4	5+ 5
6	7	8	9	10
5+ 6	5+ 7	5+ 8	5+ 9	
11	12	13	14	

6+ 1	6+ 2	6+ 3	6+ 4	6+ 5
7	8	9	10	11
6+ 6	6+ 7	6+ 8	6+ 9	
12	13	14	15	

7+ 1	7+ 2	7+ 3	7+ 4	7+ 5
8	9	10	11	12
7+ 6	7+ 7	7+ 8	7+ 9	
13	14	15	16	

8+ 1	8+ 2	8+ 3	8+ 4	8+ 5
9	10	11	12	13
8+ 6	8+ 7	8+ 8	8+ 9	
14	15	16	17	

9+ 1	9+ 2	9+ 3	9+ 4	9+ 5
10	11	12	13	14
9+ 6	9+ 7	9+ 8	9+ 9	
15	16	17	18	

CHAPTER-2
Multiplication Table-11

1 x 11 = 11	11 22
2 x 11 = 22	
3 x 11 = 33	33 44
4 x 11 = 44	
5 x 11 = 55	55 66
6 x 11 = 66	
7 x 11 = 77	77 88
8 x 11 = 88	
9 x 11 = 99	99 110
10 x 11 = 110	

Multiplication Table-11

Rule - When any single digit number is multiplied by 11, the answer is the same number repeated twice.

For example

N x 11 = N N

1 x 11 = __ __ = 1 1

2 x 11 = __ __ = 2 2

3 x 11 = __ __ = 3 3

4 x 11 = __ __ = 4 4

5 x 11 = __ __ = 5 5

6 x 1 1 = __ __ = 6 6

7 x 11 = __ __ = 7 7

8 x 11 = __ __ = 8 8

9 x 11 = __ __ = 9 9

10 x 11 = __ __ = 11 0 [Just attach zero to 11 - When any number is multiplied by 10, just attach zero to the number]

CHAPTER-3
Multiplication Table-12

1 x 12 = 12	
2 x 12 = 24	12 24
3 x 12 = 36	
4 x 12 = 48	36 48
5 x 12 = 60	
6 x 12 = 72	60 72
7 x 12 = 84	
8 x 12 = 96	84 96
9 x 12 = 108	
10 x 12 = 120	108 120

Multiplication Table–12

1 x 12 = ___ ___ [just 12 is the answer]

1 x 12 = <u>1</u> ___ [The left part of the answer is the number itself]

1 x 12 = <u>1</u> <u>2</u> [The right part of the answer is 1 x 2 = 2]

2 x 12 = ___ ___

2 x 12 = <u>2</u> ___ [The left part of the answer is the number itself]

2 x 12 = <u>2</u> <u>4</u> [The right part of the answer is 2 x 2 = 4]

3 x 12 = ___ ___

3 x 12 = 3 ___ [The left part of the answer is the number itself]

3 x 12 = 3 6 [The right part of the answer is 3 x 2 = 6]

4 x 12 = ___ ___

4 x 12 = 4 ___ [The left part of the answer is the number itself]

4 x 12 = 4 8 [The right part of the answer is 4 x 2 = 8]

5 x 12 = __ __

5 x 12 = 5 __ [The left part of the answer is the number itself]

5 x 12 = 5 ^{1}0 [The right part of the answer is 5 x 2 = 10]

5 x 12 = 5 ^{1}0 = 6 0 [Since 10 is a double digit number, the 1 should be carried to the left part of the answer. i.e. 5+1 = 6]

6 x 12 = __ __

6 x 12 = 6 __ [The left part of the answer is the number itself]

6 x 12 = 6 ^{1}2 [The right part of the answer is 6 x 2 = 12]

6 x 12 = 6 ^{1}2 = 7 2 [Since 12 is a double digit number, the 1 should be carried to the left part of the answer. i.e. 6 +1 = 7]

7 x 12 = __ __

7 x 12 = **7** __ [The left part of the answer is the number itself]

7 x 12 = **7 ^{1}4** [The right part of the answer is 7 x 2 = 14]

7 x 12 = **7 ^{1}4** = **8 4** [Since 14 is a double digit number, the 1 should be carried to the left part of the answer. i.e. 7+1 = 8]

8 x 12 = __ __

8 x 12 = **8** __ [The left part of the answer is the number itself]

8 x 12 = **8 ^{1}6** [The right part of the answer is 8 x 2 = 16]

8 x 12 = **8 ^{1}6** = **9 6** [Since 10 is a double digit number, the 1 should be carried to the left part of the answer. i.e. 8+1 = 9]

9 x 12 = __ __

9 x 12 = 9 ___ [The left part of the answer is the number itself]

9 x 12 = 9 ^{1}8 [The right part of the answer is 9 x 2 = 18]

9 x 12 = 9 ^{1}6 = 10 8 [Since 18 is a double digit number, the 1 should be carried to the left part of the answer. i.e. 9+1 = 10]

10 x 12 = 12 0 [Just attach zero to 12]

CHAPTER-4
Multiplication Table-13

1 x 13 = 13	13 26
2 x 13 = 26	
3 x 13 = 39	39 52
4 x 13 = 52	
5 x 13 = 65	65 78
6 x 13 = 78	
7 x 13 = 91	91 104
8 x 13 = 104	
9 x 13 = 117	117 130
10 x 13 = 130	

Multiplication Table–13

1 x 13 = __ __ [just 13 is the answer]

1 x 13 = 1 ___ [The left part of the answer is the number itself]

1 x 13 = 1 3 [The right part of the answer is 1 x 3 = 3]

2 x 13 = __ __

2 x 13 = 2 ___ [The left part of the answer is the number itself]

2 x 13 = 2 6 [The right part of the answer is 2 x 3 = 6]

3 x 13 = __ __

3 x 13 = 3 __ [The left part of the answer is the number itself]

3 x 13 = 3 9 [The right part of the answer is 3 x 3 = 9]

4 x 13 = __ __

4 x 13 = 4 __ [The left part of the answer is the number itself]

4 x 13 = 4 ^{1}2 [The right part of the answer is 4 x 3 = 12]

4 x 13 = 4 ^{1}2 = 5 2 [Since 12 is a double digit number, the 1 should be carried to the left part of the answer. i.e. 4+1 = 5]

5 x 13 = __ __

5 x 13 = 5 ___ [The left part of the answer is the number itself]

5 x 13 = 5 ^{1}5 [The right part of the answer is 5 x 3 = 15]

5 x 13 = 5 ^{1}5 = 6 5 [Since 15 is a double digit number, the 1 should be carried to the left part of the answer. i.e. 5+1 = 6]

6 x 13 = __ __

6 x 13 = 6 ___ [The left part of the answer is the number itself]

6 x 13 = 6 ^{1}8 [The right part of the answer is 6 x 3 = 18]

6 x 13 = 6 ^{1}8 = 7 8 [Since 18 is a double digit number, the 1 should be carried to the left part of the answer. i.e. 6 +1 = 7]

7 x 13 = __ __

7 x 13 = 7 __ [The left part of the answer is the number itself]

7 x 13 = 7 ^{2}1 [The right part of the answer is 7 x 3 = 21]

7 x 13 = 7 ^{2}1 = 9 1 [Since 21 is a double digit number, the 2 should be carried to the left part of the answer. i.e. 7+2 = 9]

8 x 13 = __ __

8 x 13 = 8 __ [The left part of the answer is the number itself]

8 x 13 = 8 ^{2}4 [The right part of the answer is 8 x 3 = 24]

8 x 13 = 8 ^{2}4 = 10 4 [Since 24 is a double digit number, the 2 should be carried to the left part of the answer. i.e. 8+2 = 10]

9 x 13 = __ __

9 x 13 = 9 ___ [The left part of the answer is the number itself]

9 x 13 = 9 ^{2}7 [The right part of the answer is 9 x 3 = 27]

9 x 13 = 9 ^{2}7 = 11 7 [Since 27 is a double digit number, the 2 should be carried to the left part of the answer. i.e. 9+2 = 11]

10 x 13 = 13 0 [Just attach zero to 13]

CHAPTER-5
Multiplication Table-14

1 x 14 = 14	
2 x 14 = 28	14 28
3 x 14 = 42	42 56
4 x 14 = 56	70 84
5 x 14 = 70	98 112
6 x 14 = 84	126 140
7 x 14 = 98	
8 x 14 = 112	
9 x 14 = 126	
10 x 14 = 140	

Multiplication Table–14

1 x 14 = __ __ [Just 14 is the answer]

1 x 14 = 1 __ [The left part of the answer is the number itself]

1 x 14 = 1 4 [The right part of the answer is 1 x 4 = 4]

2 x 14 = __ __

2 x 14 = 2 __ [The left part of the answer is the number itself]

2 x 14 = 2 8 [The right part of the answer is 2 x 4 = 8]

3 x 14 = __ __

3 x 14 = 3 __ [The left part of the answer is the number itself]

3 x 14 = 3 ^{1}2 [The right part of the answer is 3 x 4 = 12]

3 x 14 = 3 ^{1}2 = 4 2 [Since 12 is a double digit number, the 1 should be carried to the left part of the answer. i.e. 3+1 = 4]

4 x 14 = __ __

4 x 14 = 4 __ [The left part of the answer is the number itself]

4 x 14 = 4 ^{1}6 [The right part of the answer is 4 x 4 = 16]

4 x 14 = 4 ^{1}6 = 5 6 [Since 16 is a double digit number, the 1 should be carried to the left part of the answer. i.e. 4+1 = 5]

5 x 14 = __ __

5 x 14 = 5 __ [The left part of the answer is the number itself]

5 x 14 = 5 ^{2}0 [The right part of the answer is 5 x 4 = 20]

5 x 14 = 5 ^{2}0 = 7 0 [Since 20 is a double digit number, the 2 should be carried to the left part of the answer. i.e. 5 + 2 = 7]

6 x 14 = __ __

6 x 14 = 6 __ [The left part of the answer is the number itself]

6 x 14 = 6 ^{2}4 [The right part of the answer is 6 x 4 = 24]

6 x 14 = 6 ^{2}4 = 8 4 [Since 24 is a double digit number, the 2 should be carried to the left part of the answer. i.e. 6 + 2 = 8]

7 x 14 = __ __

7 x 14 = 7 __   [The left part of the answer is the number itself]

7 x 14 = 7 ^{2}8   [The right part of the answer is 7 x 4 = 28]

7 x 14 = 7 ^{2}8 = 9 8 [Since 28 is a double digit number, the 2 should be carried to the left part of the answer. i.e. 7+2 = 9]

8 x 14 = __ __

8 x 14 = 8 __   [The left part of the answer is the number itself]

8 x 14 = 8 ^{3}2   [The right part of the answer is 8 x 4 = 32]

8 x 14 = 8 ^{3}2 = 11 2 [Since 32 is a double digit number, the 3 should be carried to the left part of the answer. i.e. 8+3 = 11]

9 x 14 = __ __

9 x 14 = 9 ___ [The left part of the answer is the number itself]

9 x 14 = 9 ^{3}6 [The right part of the answer is 9 x 4 = 36]

9 x 14 = 9 ^{3}6 = 12 6 [Since 36 is a double digit number, the 3 should be carried to the left part of the answer. i.e. 9+3 = 12]

10 x 14 = 14 0 [Just attach zero to 14]

CHAPTER-6

Multiplication Table-15

1 x 15 = 15	
2 x 15 = 30	15 30
3 x 15 = 45	45 60
4 x 15 = 60	
5 x 15 = 75	75 90
6 x 15 = 90	
7 x 15 = 105	105 120
8 x 15 = 120	
9 x 15 = 135	135 150
10 x 15 = 150	

Multiplication Table–15

1 x 15 = __ __ [Just 15 is the answer]

1 x 15 = 1 __ [The left part of the answer is the number itself]

1 x 15 = 1 5 [The right part of the answer is 1 x 5 = 5]

2 x 15 = __ __

2 x 15 = 2 __ [The left part of the answer is the number itself]

2 x 15 = 2 ^{1}0 [The right part of the answer is 2 x 5 = 10]

2 x 15 = 2 ^{1}0 = 3 0 [Since 10 is a double digit number, the 1 should be carried to the left part of the answer. i.e. 2+1 = 3]

3 x 15 = __ __

3 x 15 = 3 __ [The left part of the answer is the number itself]

3 x 15 = 3 ^{1}5 [The right part of the answer is 3 x 5 = 15]

3 x 14 = 3 ^{1}2 = 4 5 [Since 15 is a double digit number, the 1 should be carried to the left part of the answer. i.e. 3+1 = 4]

4 x 15 = __ __

4 x 15 = 4 __ [The left part of the answer is the number itself]

4 x 15 = 4 ^{2}0 [The right part of the answer is 4 x 5 = 20]

4 x 15 = 4 ^{2}0 = 6 0 [Since 20 is a double digit number, the 2 should be carried to the left part of the answer. i.e. 4+2 = 6]

5 x 15 = __ __

5 x 15 = 5 ___ [The left part of the answer is the number itself]

5 x 15 = 5 ^{2}5 [The right part of the answer is 5 x 5 = 25]

5 x 15 = 5 ^{2}5 = 7 5 [Since 25 is a double digit number, the
2 should be carried to the left part of the answer. i.e. 5 + 2 = 7]

6 x 15 = __ __

6 x 15 = 6 ___ [The left part of the answer is the number itself]

6 x 15 = 6 ^{3}0 [The right part of the answer is 6 x 5 = 30]

6 x 15 = 6 ^{3}0 = 9 0 [Since 30 is a double digit number, the
2 should be carried to the left part of the answer. i.e. 6 + 3 = 9]

7 x 15 = __ __

7 x 15 = 7 ___ [The left part of the answer is the number itself]

7 x 15 = 7 ^{3}5 [The right part of the answer is 7 x 5 = 35]

7 x 15 = 7 ^{3}5 = 10 5 [Since 35 is a double digit number, the 3 should be carried to the left part of the answer. i.e. 7+3 = 10]

8 x 15 = __ __

8 x 15 = 8 ___ [The left part of the answer is the number itself]

8 x 15 = 8 ^{4}0 [The right part of the answer is 8 x 5 = 40]

8 x 15 = 8 ^{4}0 = 12 0 [Since 40 is a double digit number, the 4 should be carried to the left part of the answer. i.e. 8+4 = 12]

9 x 15 = __ __

9 x 15 = 9 ___ [The left part of the answer is the number itself]

9 x 15 = 9 ^{4}5 [The right part of the answer is 9 x 5 = 45]

9 x 15 = 9 ^{4}5 = 13 5 [Since 45 is a double digit number, the 4 should be carried to the left part of the answer. i.e. 9+4 = 13]

10 x 15 = 15 0 [Just attach zero to 15]

CHAPTER-7
Multiplication Table-16

1 x 16 = 16	
2 x 16 = 32	16 32
3 x 16 = 48	48 64
4 x 16 = 64	80 96
5 x 16 = 80	112 128
6 x 16 = 96	144 160
7 x 16 = 112	
8 x 16 = 128	
9 x 16 = 144	
10 x 16 = 160	

Multiplication Table–16

1 x 16 = ___ ___ [Just 16 is the answer]

1 x 16 = 1 ___ [The left part of the answer is the number itself]

1 x 16 = 1 6 [The right part of the answer is 1 x 6 = 6]

2 x 16 = ___ ___

2 x 16 = 2 ___ [The left part of the answer is the number itself]

2 x 16 = 2 ^{1}2 [The right part of the answer is 2 x 6 = 12]

2 x 16 = 2 ^{1}2 = 3 2 [Since 12 is a double digit number, the 1 should be carried to the left part of the answer. i.e. 2+1 = 3]

3 x 16 = __ __

3 x 16 = 3 __   [The left part of the answer is the number itself]

3 x 16 = $3\,^{1}8$ [The right part of the answer is 3 x 6 = 18]

3 x 16 = $3\,^{1}8$ = 4 8 [Since 18 is a double digit number, the 1 should be carried to the left part of the answer. i.e. 3+1 = 4]

4 x 16 = __ __

4 x 16 = 4 __   [The left part of the answer is the number itself]

4 x 16 = $4\,^{2}4$ [The right part of the answer is 4 x 6 = 24]

4 x 16 = $4\,^{2}4$ = 6 4 [Since 24 is a double digit number, the 1 should be carried to the left part of the answer. i.e. 4+2 = 6]

5 x 16 = __ __

5 x 16 = 5 __   [The left part of the answer is the number itself]

5 x 16 = 5 ^{3}0   [The right part of the answer is 5 x 6 = 30]

5 x 16 = 5 ^{3}0 = 8 0 [Since 30 is a double digit number, the 3 should be carried to the left part of the answer. i.e. 5+3 = 8]

6 x 16 = __ __

6 x 16 = 6 __   [The left part of the answer is the number itself]

6 x 16 = 6 ^{3}6   [The right part of the answer is 6 x 6 = 36]

6 x 16 = 6 ^{3}6 = 9 6 [Since 36 is a double digit number, the 3 should be carried to the left part of the answer. i.e. 6 + 3 = 9]

7 x 16 = __ __

7 x 16 = 7 ___ [The left part of the answer is the number itself]

7 x 16 = 7 ^{4}2 [The right part of the answer is 7 x 6 = 42]

7 x 16 = 7 ^{4}2 = 11 2 [Since 42 is a double digit number, the 4 should be carried to the left part of the answer. i.e. 7+4 = 11]

8 x 16 = __ __

8 x 16 = 8 ___ [The left part of the answer is the number itself]

8 x 16 = 8 ^{4}8 [The right part of the answer is 8 x 6 = 48]

8 x 16 = 8 ^{4}8 = 12 8 [Since 48 is a double digit number, the 4 should be carried to the left part of the answer. i.e. 8+4 = 12]

9 x 16 = __ __ =

9 x 16 = 9 ___ [The left part of the answer is the number itself]

9 x 16 = 9 ^{5}4 [The right part of the answer is 9 x 6 = 54]

9 x 16 = 9 ^{5}4 = 14 4 [Since 54 is a double digit number, the 5 should be carried to the left part of the answer. i.e. 9+5 = 14]

10 x 16 = 16 0 [Just attach zero to 16]

CHAPTER-8

Multiplication Table-17

1 x 17 = 17	
2 x 17 = 34	
3 x 17 = 51	17 34
4 x 17 = 68	51 68
5 x 17 = 85	
6 x 17 = 102	85 102
7 x 17 = 119	119 136
8 x 17 = 136	
9 x 17 = 153	153 170
10 x 17 = 170	

Multiplication Table–17

1 x 17 = __ __ [Just 17 is the answer]

1 x 17 = 1 __ [The left part of the answer is the number itself]

1 x 17 = 1 7 [The right part of the answer is 1 x 7 = 7]

2 x 17 = __ __

2 x 17 = 2 __ [The left part of the answer is the number itself]

2 x 17 = 2 ^{1}4 [The right part of the answer is 2 x 7 = 14]

2 x 17 = 2 ^{1}4 = 3 4 [Since 14 is a double digit number, the 1 should be carried to the left part of the answer. i.e. 2+1 = 3]

3 x 17 = __ __

3 x 17 = 3 __ [The left part of the answer is the number itself]

3 x 17 = 3 ^{2}1 [The right part of the answer is 3 x 7 = 21]

3 x 17 = 3 ^{2}1 = 5 1 [Since 21 is a double digit number, the 2 should be carried to the left part of the answer. i.e. 3+2 = 5]

4 x 17 = __ __

4 x 17 = 4 __ [The left part of the answer is the number itself]

4 x 17 = 4 ^{2}8 [The right part of the answer is 4 x 7 = 28]

4 x 17 = 4 ^{2}8 = 6 8 [Since 28 is a double digit number, the 2 should be carried to the left part of the answer. i.e. 4+2 = 6]

5 x 17 = __ __

5 x 17 = 5 ___ [The left part of the answer is the number itself]

5 x 17 = 5 ^{3}5 [The right part of the answer is 5 x 7 = 35]

5 x 17 = 5 ^{3}5 = 8 5 [Since 35 is a double digit number, the 3 should be carried to the left part of the answer. i.e. 5+3 = 8]

6 x 17 = __ __

6 x 17 = 6 ___ [The left part of the answer is the number itself]

6 x 17 = 6 ^{4}2 [The right part of the answer is 6 x 7 = 42]

6 x 17 = 6 ^{4}2 = 10 2 [Since 42 is a double digit number, the 4 should be carried to the left part of the answer. i.e. 6+4 = 10]

7 x 17 = __ __

7 x 17 = 7 ___ [The left part of the answer is the number itself]

7 x 17 = 7 ^{4}9 [The right part of the answer is 7 x 7 = 49]

7 x 17 = 7 ^{4}9 = 11 9 [Since 49 is a double digit number, the 4 should be carried to the left part of the answer. i.e. 7+4 = 11]

8 x 17 = __ __

8 x 17 = 8 ___ [The left part of the answer is the number itself]

8 x 17 = 8 ^{5}6 [The right part of the answer is 8 x 7 = 56]

8 x 17 = 8 ^{5}6 = 13 6 [Since 56 is a double digit number, the 5 should be carried to the left part of the answer. i.e. 8+5 = 13]

9 x 17 = __ __

9 x 17 = 9 ___ [The left part of the answer is the number itself]

9 x 17 = 9 ^{6}3 [The right part of the answer is 9 x 7 = 63]

9 x 17 = 9 ^{6}3 = 15 3 [Since 63 is a double digit number, the 6 should be carried to the left part of the answer. i.e. 9+6 = 15]

10 x 17 = 17 0 [Just attach zero to 17]

CHAPTER-9

Multiplication Table-18

1 x 18 = 18	
2 x 18 = 36	
3 x 18 = 54	18 36
4 x 18 = 72	54 72
5 x 18 = 90	
6 x 18 = 108	90 108
7 x 18 = 126	126 144
8 x 18 = 144	
9 x 18 = 162	162 180
10 x 18 = 180	

Multiplication Table–18

1 x 18 = __ __ [Just 18 is the answer]

1 x 18 = 1 __ [The left part of the answer is the number itself]

1 x 18 = 1 8 [The right part of the answer is 1 x 8 = 8]

2 x 18 = __ __

2 x 18 = 2 __ [The left part of the answer is the number itself]

2 x 18 = 2 ^{1}6 [The right part of the answer is 2 x 8 = 16]

2 x 18 = 2 ^{1}6 = 3 6 [Since 16 is a double digit number, the 1 should be carried to the left part of the answer. i.e. 2+1 = 3]

3 x 18 = __ __

3 x 18 = 3 __ [The left part of the answer is the number itself]

3 x 18 = 3 ^{2}4 [The right part of the answer is 3 x 8 = 24]

3 x 18 = 3 ^{2}4 = 5 4 [Since 24 is a double digit number, the 2 should be carried to the left part of the answer. i.e. 3+2 = 5]

4 x 18 = __ __

4 x 18 = 4 __ [The left part of the answer is the number itself]

4 x 18 = 4 ^{3}2 [The right part of the answer is 4 x 8 = 32]

4 x 18 = 4 ^{3}2 = 7 2 [Since 32 is a double digit number, the 3 should be carried to the left part of the answer. i.e. 4+3 = 7]

5 x 18 = ___ ___

5 x 18 = 5 ___ [The left part of the answer is the number itself]

5 x 18 = 5 ^{4}0 [The right part of the answer is 5 x 8 = 40]

5 x 18 = 5 ^{4}0 = 9 0 [Since 40 is a double digit number, the 4 should be carried to the left part of the answer. i.e. 5+4 = 9]

6 x 18 = ___ ___

6 x 18 = 6 ___ [The left part of the answer is the number itself]

6 x 18 = 6 ^{4}8 [The right part of the answer is 6 x 8 = 48]

6 x 18 = 6 ^{4}8 = 10 8 [Since 48 is a double digit number, the 4 should be carried to the left part of the answer. i.e. 6+4 = 10]

7 x 18 = __ __

7 x 18 = 7 __ [The left part of the answer is the number itself]

7 x 18 = 7 ^{5}6 [The right part of the answer is 7 x 8 = 56]

7 x 18 = 7 ^{5}6 = 12 6 [Since 56 is a double digit number, the 5 should be carried to the left part of the answer. i.e. 7+5 = 12]

8 x 18 = __ __

8 x 18 = 8 __ [The left part of the answer is the number itself]

8 x 18 = 8 ^{6}4 [The right part of the answer is 8 x 8 = 64]

8 x 18 = 8 ^{6}4 = 14 4 [Since 64 is a double digit number, the 6 should be carried to the left part of the answer. i.e. 8+6 = 14]

9 x 18 = __ __

9 x 18 = 9 __ [The left part of the answer is the number itself]

9 x 18 = 9 ^{7}2 [The right part of the answer is 9 x 8 = 72]

9 x 18 = 9 ^{7}2 = 16 2 [Since 72 is a double digit number, the 7 should be carried to the left part of the answer. i.e. 9+7 = 16]

10 x 18 = 18 0 [Just attach zero to 18]

CHAPTER-10

Multiplication Table-19

1 x 19 = 19	
2 x 19 = 38	
3 x 19 = 57	19 38
4 x 19 = 76	57 76
5 x 19 = 95	
6 x 19 = 114	95 114
7 x 19 = 133	133 152
8 x 19 = 152	
9 x 19 = 171	171 190
10 x 19 = 190	

Multiplication Table–19

1 x 19 = __ __ [Just 19 is the answer]

1 x 19 = 1 ___ [The left part of the answer is the number itself]

1 x 19 = 1 9 [The right part of the answer is 1 x 9 = 9]

2 x 19 = __ __

2 x 19 = 2 ___ [The left part of the answer is the number itself]

2 x 19 = 2 ^{1}8 [The right part of the answer is 2 x 9 = 18]

2 x 19 = 2 ^{1}8 = 3 8 [Since 18 is a double digit number, the 1 should be carried to the left part of the answer. i.e. 2+1 = 3]

3 x 19 = __ __

3 x 19 = 3 __ [The left part of the answer is the number itself]

3 x 19 = 3 ^{2}7 [The right part of the answer is 3 x 9 = 27]

3 x 19 = 3 ^{2}7 = 5 7 [Since 27 is a double digit number, the 2 should be carried to the left part of the answer. i.e. 3+2 = 5]

4 x 19 = __ __

4 x 19 = 4 __ [The left part of the answer is the number itself]

4 x 19 = 4 ^{3}6 [The right part of the answer is 4 x 9 = 36]

4 x 19 = 4 ^{3}6 = 7 6 [Since 36 is a double digit number, the 3 should be carried to the left part of the answer. i.e. 4+3 = 7]

5 x 19 = __ __

5 x 19 = 5 __ [The left part of the answer is the number itself]

5 x 19 = 5 ^{4}5 [The right part of the answer is 5 x 9 = 45]

5 x 19 = 5 ^{4}5 = 9 5 [Since 45 is a double digit number, the 4 should be carried to the left part of the answer. i.e. 5+4 = 9]

6 x 19 = __ __

6 x 19 = 6 __ [The left part of the answer is the number itself]

6 x 19 = 6 ^{5}4 [The right part of the answer is 6 x 9 = 54]

6 x 19 = 6 ^{5}4 = 11 4 [Since 54 is a double digit number, the 5 should be carried to the left part of the answer. i.e. 6+5 = 11]

7 x 19 = __ __

7 x 19 = 7 __   [The left part of the answer is the number itself]

7 x 19 = 7 ^{6}3   [The right part of the answer is 7 x 9 = 63]

7 x 19 = 7 ^{6}3 = 13 3 [Since 63 is a double digit number, the 6 should be carried to the left part of the answer. i.e. 7+6 = 13]

8 x 19 = __ __

8 x 19 = 8 __   [The left part of the answer is the number itself]

8 x 19 = 8 ^{7}2   [The right part of the answer is 8 x 9 = 72]

8 x 19 = 8 ^{7}2 = 15 2 [Since 72 is a double digit number, the 7 should be carried to the left part of the answer. i.e. 8+7 = 15]

9 x 19 = __ __

9 x 19 = 9 ___ [The left part of the answer is the number itself]

9 x 19 = 9 ^{8}1 [The right part of the answer is 9 x 9 = 81]

9 x 19 = 9 ^{8}1 = 17 1 [Since 81 is a double digit number, the 8 should be carried to the left part of the answer. i.e. 9+8 = 17]

10 x 19 = 19 0 [Just attach zero to 19]

CHAPTER-11
Multiplication Table-20

1 x 20 = 20	
2 x 20 = 40	
3 x 20 = 60	20 40
4 x 20 = 80	
5 x 20 = 100	60 80
6 x 20 = 120	100 120
7 x 20 = 140	
8 x 20 = 160	140 160
9 x 20 = 180	180 200
10 x 20 = 200	

Multiplication Table–20

1 x 20 = ___ ___ [Just 20 is the answer]

1 x 20 = 2 ___ [The left part of the answer is 1 x 2 = 2]

1 x 20 = <u>2</u> <u>0</u> [The right part of the answer is zero]

2 x 20 = ___ ___

2 x 20 = 4 ___ [The left part of the answer is 2 x 2 = 4]

2 x 20 = <u>4</u> <u>0</u> [The right part of the answer is zero]

3 x 20 = __ __

3 x 20 = 6 __ [The left part of the answer is 3 x 2 = 6]

3 x 20 = 6 0 [The right part of the answer is zero]

4 x 20 = __ __

4 x 20 = 8 __ [The left part of the answer is 4 x 2 = 8]

4 x 20 = 8 0 [The right part of the answer is zero]

5 x 20 = __ __

5 x 20 = 10 __ [The left part of the answer is 5 x 2 = 10]

5 x 20 = <u>10</u> <u>0</u> [The right part of the answer is zero]

6 x 20 = __ __

6 x 20 = 12 __ [The left part of the answer is 6 x 2 = 12]

6 x 20 = <u>12</u> <u>0</u> [The right part of the answer is zero]

7 x 20 = __ __

7 x 20 = 14 __   [The left part of the answer is 7 x 2 = 14]

7 x 20 = <u>14</u> <u>0</u>   [The right part of the answer is zero]

8 x 20 = __ __

8 x 20 = 16 __   [The left part of the answer is 8 x 2 = 16]

8 x 20 = <u>16</u> <u>0</u>   [The right part of the answer is zero]

9 x 20 = __ __

9 x 20 = 18 __ [The left part of the answer is 9 x 2 = 18]

9 x 20 = <u>18</u> <u>0</u> [The right part of the answer is zero]

10 x 20 = <u>20</u> <u>0</u> [Just attach zero to 20]

Conclusion

Of course, the rules are not really conquered until they can be used without thinking of them. A little practice makes the rules sufficiently automatic to accomplish this. Practicing these methods for an hour is actually very little when we compare it to the hours and hours of repeated drill that children in primary schools devote to learning all the facts of multiplication tables.

www.ingramcontent.com/pod-product-compliance
Lightning Source LLC
Chambersburg PA
CBHW081537220526
45467CB00010B/3236